Henry Pickering Bowditch

The Growth of Children

Studied by Galton's Method of Percentile Grades

Henry Pickering Bowditch

The Growth of Children
Studied by Galton's Method of Percentile Grades

ISBN/EAN: 9783337778842

Printed in Europe, USA, Canada, Australia, Japan

Cover: Foto ©berggeist007 / pixelio.de

More available books at **www.hansebooks.com**

THE GROWTH OF CHILDREN,

STUDIED BY

GALTON'S METHOD OF PERCENTILE GRADES,

By H. P. BOWDITCH, M.D.,

PROFESSOR OF PHYSIOLOGY, HARVARD MEDICAL SCHOOL.

Reprinted from the Twenty-second Annual Report of the State Board of Health of Massachusetts. Boston, 1891.

THE GROWTH OF CHILDREN,

GALTON'S METHOD OF PERCENTILE GRADES,

BY H. P. BOWDITCH, M.D.,

Professor of Physiology, Harvard Medical School.

In the last report of the Massachusetts State Board of Health the advantages of discussing statistical data by Galton's method [*] of percentile grades were explained and illustrated in a paper entitled " The Physique of Women in Massachusetts." The value of the method in anthropometrical work seemed so obvious that it has been thought desirable to apply it to the large body of observations on the height and weight of Boston school children which formed the basis of an article on "The Growth of Children," published by the Board of Health in 1877.

In this article, at the suggestion of Mr. Charles Roberts, tables were given showing the *distribution* of the observations; *i.e.*, the number of individuals at each age whose height was recorded at each successive inch or whose weight fell within successive groups of four pounds each. From these tables (tables 4 to 15 of the above-mentioned article) it was easy to calculate the values at the various percentile grades. For example, it appears from Table 4 that the heights of 848 boys between five and six years old were distributed as shown in the following table : —

---

[*] See " Galton, Natural Inheritance," London, McMillan & Co., 1889.

Distribution of Observations on Heights of Boston School-boys. Age at Last Birthday Five Years.

Inches.	Number of observations.	Inches.	Number of observations.	Inches.	Number of observations.
47	4	41	190	35	1
46	8	40	149	34	-
45	20	39	79	33	-
44	62	38	42	32	-
43	119	37	17	31	-
42	149	36	7	30	1

Total number of observations, ... 848

In this table it will be seen that five per cent., for instance, of the total number of observations is 42.4. Now since the observations corresponding to each successive inch include all the measurements between that inch and the next inch above, it is evident that there are $1+1+7+17 = 26$ individuals less than 38 inches in height and $1+1+7+17+42 = 68$ individuals less than 39 inches in height. Since, therefore, 42.4 lies between 26 and 68 it follows that the height below which five per cent. of the observations fall must be between 38 and 39 inches. The exact height can readily be calculated by interpolation. Thus the fraction of an inch to be added to 38 to give the required height is obtained by dividing 16.4 (*i.e.*, 42.4 — 26) by 42 (*i.e.*, the number of observations between 38 and 39 inches). This fraction is 0.39, and, therefore, 38.39 inches is the height below which five per cent. and above which ninety-five per cent. of the observations fall ; *i.e.*, it is the value of the five per cent. grade.

In this way tables 1 to 12 have been calculated from tables 4 to 15 inclusive of the original article. These tables show the heights and weights of Boston school children of both sexes and various ages at percentile grades varying from five per cent. to ninety-five per cent. Separate tables are, moreover, given for children of American parentage, Irish parentage, and for the whole number of observations irrespective of nationality. The values are given in both the English and the metric system of weights and measures, and in the last column of each table are to be found the average heights and weights of children of each age as given in the original article.

The conclusions which may be drawn from a study of these tables will be best understood after an examination of the curves which have been constructed from them, and as a preliminary to this study it will be well to consider briefly the general character of curves representing values at various percentile grades.

TABLES SHOWING HEIGHTS AND WEIGHTS, AMERICAN, IRISH AND TOTAL, AT PERCENTILE GRADES.

TABLE No. 1.—CALCULATED FROM TABLE 4 OF ORIGINAL ARTICLE.

Heights of Boston School-boys irrespective of Nationality.

Age at Last Birthday.	Number of Observations.	Unit of Measurement.	Values at the Following Percentile Grades.											Average.
			5 Per Cent.	10 Per Cent.	20 Per Cent.	30 Per Cent.	40 Per Cent.	50 Per Cent.	60 Per Cent.	70 Per Cent.	80 Per Cent.	90 Per Cent.	95 Per Cent.	
Five,	849	inch,	38.39	39.21	40.15	40.72	41.23	41.67	42.15	42.72	43.36	44.15	44.83	41.57
		c. m.,	97.5	99.6	102.0	103.4	104.7	105.8	107.1	108.6	110.1	112.1	113.9	105.6
Six,	1,258	inch,	40.66	41.43	42.31	42.80	43.30	43.87	44.37	44.86	45.54	46.40	47.13	43.75
		c. m.,	103.3	105.2	107.4	108.9	110.2	111.4	112.7	113.9	115.7	118.1	119.7	111.1
Seven,	1,410	inch,	42.48	43.27	44.20	44.76	45.31	45.83	46.36	46.88	47.65	48.48	49.27	45.74
		c. m.,	107.9	109.9	112.3	113.7	115.1	116.4	117.8	119.1	120.8	123.1	125.1	116.2
Eight,	1,481	inch,	44.46	45.24	46.15	46.81	47.35	47.84	48.36	48.90	49.61	50.58	51.42	47.76
		c. m.,	112.9	114.9	117.2	118.9	120.3	121.5	122.8	124.2	126.0	128.5	130.6	121.3
Nine,	1,437	inch,	46.41	47.18	48.04	48.70	49.27	49.77	50.30	50.87	51.60	52.61	53.40	49.00
		c. m.,	117.9	119.8	122.0	123.7	125.1	126.4	127.8	129.2	131.1	133.6	135.8	124.5
Ten,	1,363	inch,	48.11	48.98	49.77	50.45	51.10	51.73	52.34	52.94	53.76	54.85	55.83	51.68
		c. m.,	122.2	124.4	126.4	128.1	129.8	131.4	132.9	134.5	136.6	139.3	141.8	131.3
Eleven,	1,293	inch,	49.47	50.32	51.30	52.08	52.75	53.40	54.04	54.73	55.50	56.54	57.50	53.33
		c. m.,	125.7	127.8	130.3	132.3	134.0	135.6	137.3	139.0	141.0	143.6	146.0	135.4
Twelve,	1,253	inch,	51.10	51.96	53.07	53.78	54.46	55.11	55.75	56.46	57.36	58.75	59.88	55.11
		c. m.,	129.8	132.0	134.8	136.6	138.3	140.0	141.6	143.4	145.7	149.2	152.1	140.0
Thirteen,	1,160	inch,	52.62	53.58	54.81	55.60	56.37	56.98	57.84	58.75	59.80	61.47	62.73	57.21
		c. m.,	133.7	136.1	139.2	141.5	143.2	144.7	146.9	149.2	151.9	156.1	159.3	145.3
Fourteen,	908	inch,	54.57	55.66	57.13	58.07	58.78	59.60	60.48	61.51	62.80	64.61	65.80	59.88
		c. m.,	138.6	141.4	145.1	147.5	149.3	151.4	153.6	156.2	159.5	164.1	167.4	152.1
Fifteen,	636	inch,	56.55	58.07	59.46	60.30	61.33	62.37	63.35	64.41	65.48	66.85	67.90	62.30
		c. m.,	143.6	147.5	151.0	153.4	155.8	158.4	160.9	163.6	166.3	169.8	172.5	158.2
Sixteen,	350	inch,	59.76	61.21	62.90	63.76	64.68	65.35	66.06	66.74	67.47	68.65	69.86	66.00
		c. m.,	151.8	155.5	159.5	161.9	164.0	166.0	167.8	169.5	171.4	174.4	177.4	165.1
Seventeen,	192	inch,	61.40	62.60	63.85	64.96	65.57	66.21	67.01	67.97	68.74	69.98	70.94	66.16
		c. m.,	156.0	159.0	162.2	165.0	166.5	168.2	170.2	172.6	174.6	177.4	180.2	168.0
Eighteen,	84	inch,	63.04	63.58	64.87	65.55	66.19	66.79	67.30	67.90	68.75	70.10	70.80	66.66
		c. m.,	160.1	162.3	164.8	166.5	168.1	169.6	171.2	172.7	174.6	178.1	179.8	169.3

TABLE No. 2. — CALCULATED FROM TABLE 5 OF ORIGINAL ARTICLE.

Heights of Boston School-boys of American Parentage.

| Age at Last Birthday | Number of Observations | Unit of Measurement | Values at the following percentile grades. | | | | | | | | | | | Average. |
			5 Per Cent.	10 Per Cent.	20 Per Cent.	30 Per Cent.	40 Per Cent.	50 Per Cent.	60 Per Cent.	70 Per Cent.	80 Per Cent.	90 Per Cent.	95 Per Cent.	
Five,	201	inch,	38.22	39.26	40.34	40.95	41.40	41.84	42.24	42.88	43.54	44.38	44.94	41.74
		c.m.,	97.06	99.72	102.46	104.01	105.16	106.27	107.54	108.91	110.59	112.72	114.15	106.0
Six,	342	inch,	41.09	41.83	42.56	43.14	43.57	44.00	44.57	45.17	45.87	46.79	47.62	44.10
		c.m.,	104.37	106.25	108.10	109.58	110.67	111.76	113.21	114.73	116.51	118.85	120.95	112.0
Seven,	369	inch,	42.79	43.83	44.66	45.44	45.94	46.36	46.87	47.41	47.96	48.92	49.70	46.21
		c.m.,	108.69	111.33	113.41	115.42	116.44	117.76	119.06	120.43	121.83	124.27	126.25	117.4
Eight,	407	inch,	44.57	45.45	46.43	47.18	47.78	48.34	48.88	49.46	50.06	50.98	51.93	48.16
		c.m.,	113.21	115.44	117.93	119.83	121.36	122.78	124.15	125.62	127.15	129.49	131.91	122.3
Nine,	381	inch,	46.92	47.44	48.30	49.04	49.55	50.07	50.64	51.25	51.96	53.09	54.00	50.09
		c.m.,	119.18	120.50	122.68	124.56	125.86	127.18	128.62	130.17	131.98	134.85	137.16	127.2
Ten,	360	inch,	48.91	49.39	50.20	51.02	51.65	52.24	52.76	53.44	54.32	55.50	56.33	52.21
		c.m.,	124.24	125.45	127.51	129.59	131.19	132.69	134.01	135.74	137.97	140.97	143.08	132.6
Eleven,	350	inch,	50.06	50.76	52.02	52.68	53.39	54.14	54.86	55.62	56.37	57.26	58.29	54.01
		c.m.,	127.15	128.93	132.13	133.80	135.61	137.51	139.34	141.27	143.18	145.44	148.06	137.2
Twelve,	373	inch,	51.48	52.35	53.54	54.30	55.07	55.68	56.37	57.25	58.43	59.68	60.71	55.78
		c.m.,	130.76	132.97	135.99	138.07	139.88	141.43	143.18	145.41	148.41	151.59	154.20	141.7
Thirteen,	301	inch,	53.32	54.23	55.09	56.52	57.26	58.14	58.95	59.80	60.93	62.38	63.65	58.17
		c.m.,	135.43	137.74	141.45	143.56	145.44	147.68	149.73	151.89	154.76	158.45	161.67	147.7
Fourteen,	386	inch,	54.44	56.14	58.11	58.89	59.80	60.77	61.87	62.98	64.18	65.62	67.31	61.08
		c.m.,	138.28	142.85	147.60	149.60	151.89	154.36	157.15	159.97	163.02	166.67	170.97	155.1
Fifteen,	342	inch,	57.02	58.80	60.08	61.22	62.33	63.17	64.14	65.04	65.96	67.41	68.42	62.96
		c.m.,	144.83	149.45	152.60	155.50	158.32	160.45	162.92	165.20	167.54	171.22	173.79	159.9
Sixteen,	222	inch,	60.80	61.93	63.27	64.74	65.23	66.03	66.62	67.22	67.82	69.16	70.22	65.58
		c.m.,	154.43	157.30	160.71	163.42	165.68	167.72	169.21	170.74	172.29	175.67	178.36	166.5
Seventeen,	128	inch,	61.70	62.64	64.20	65.11	65.72	66.39	67.14	68.08	68.76	70.03	70.94	66.29
		c.m.,	156.72	159.11	163.07	165.35	166.93	168.63	170.54	172.92	174.65	178.00	180.19	168.4
Eighteen,	65	inch,	63.07	64.08	65.08	65.62	66.18	66.77	67.26	67.95	68.86	70.42	70.95	66.76
		c.m.,	160.20	162.76	165.30	166.67	168.10	169.60	171.09	172.59	174.90	178.87	180.21	169.5

TABLE No. 3.—CALCULATED FROM TABLE 6 OF ORIGINAL ARTICLE.

Heights of Boston School-boys of Irish Parentage.

Age at Last Birthday	Number of Observations	Unit of Measurement	Values at the following Percentile Grades.											Average
			5 Per Cent.	10 Per Cent.	20 Per Cent.	30 Per Cent.	40 Per Cent.	50 Per Cent.	60 Per Cent.	70 Per Cent.	80 Per Cent.	90 Per Cent.	95 Per Cent.	
Five,	366	inch,	38.66	39.31	40.13	40.70	41.23	41.70	42.21	42.77	43.41	44.17	44.76	41.59
		c. m.,	98.2	99.8	101.9	103.4	104.7	105.9	107.2	108.6	110.3	112.2	113.7	105.5
Six,	503	inch,	40.58	41.41	42.31	42.88	43.40	43.90	44.33	44.74	45.30	46.22	47.05	43.74
		c. m.,	103.1	105.2	107.5	108.9	110.2	111.5	112.6	113.6	115.1	117.4	119.5	111.1
Seven,	562	inch,	42.47	43.22	44.08	44.63	45.17	45.68	46.19	46.67	47.27	48.15	49.09	45.61
		c. m.,	107.9	109.8	112.0	113.4	114.7	116.0	117.3	118.5	120.1	122.3	124.7	115.8
Eight,	588	inch,	44.77	45.58	46.24	46.88	47.36	47.90	48.20	48.73	49.38	50.37	51.00	47.72
		c. m.,	113.7	115.3	117.4	119.1	120.3	121.4	122.6	123.8	125.4	127.9	129.8	121.2
Nine,	556	inch,	46.51	47.18	48.03	48.00	49.14	49.61	50.11	50.72	51.48	52.46	53.12	49.53
		c. m.,	118.1	119.8	122.0	123.4	124.8	126.0	127.3	128.8	130.8	133.2	134.9	125.2
Ten,	571	inch,	47.93	48.93	49.75	50.40	50.09	51.62	52.22	52.77	53.56	54.66	55.73	51.57
		c. m.,	121.7	124.3	126.4	128.0	129.5	131.1	132.6	134.0	136.0	138.8	141.5	131.1
Eleven,	548	inch,	49.41	50.34	51.29	51.98	52.58	53.17	53.76	54.42	55.16	56.01	56.06	53.10
		c. m.,	125.5	127.9	130.3	132.0	133.6	135.0	136.5	138.2	140.1	142.3	144.6	134.9
Twelve,	497	inch,	51.02	51.76	52.80	53.58	54.27	54.80	55.51	56.14	56.85	58.08	59.57	54.82
		c. m.,	129.6	131.5	134.1	136.1	137.8	139.4	141.0	142.6	144.4	147.5	151.3	139.3
Thirteen,	463	inch,	52.54	53.46	54.59	55.36	55.99	56.58	57.26	58.17	59.06	60.37	61.74	56.70
		c. m.,	133.4	135.8	138.7	140.6	142.2	143.7	145.4	147.7	150.0	153.3	156.8	144.0
Fourteen,	334	inch,	54.44	55.52	56.80	57.62	58.25	58.81	59.50	60.38	61.31	62.73	63.94	58.88
		c. m.,	138.3	141.0	144.5	146.3	147.9	149.4	151.3	153.4	155.7	159.3	162.4	149.5
Fifteen,	155	inch,	56.24	57.28	58.75	59.02	60.33	60.98	61.83	62.96	64.00	66.37	66.07	61.15
		c. m.,	142.8	145.5	149.2	151.4	153.2	154.9	157.0	159.9	162.6	166.0	167.8	155.3
Sixteen,	61	inch,	58.50	59.35	61.40	63.19	63.70	64.42	65.29	65.97	66.83	67.70	68.45	64.00
		c. m.,	148.6	150.7	156.0	160.5	161.8	163.6	165.8	167.6	169.7	172.0	173.9	162.8
Seventeen, Eighteen,	26, 5	inch,	60.50	63.10	63.64	65.06	65.68	66.75	67.52	68.17	68.96	69.07	70.70	66.20
		c. m.,	153.7	160.3	161.6	165.2	166.8	169.5	171.4	173.1	175.1	177.2	179.6	168.2

TABLE No. 4.—CALCULATED FROM TABLE 7 OF ORIGINAL ARTICLE.

Weights of Boston School-boys irrespective of Nationality.

Age at Last Birthday.	Number of Observations.	Unit of Weights.	Values at the Following Percentile Grades.											Average.
			5 Per Cent.	10 Per Cent.	20 Per Cent.	30 Per Cent.	40 Per Cent.	50 Per Cent.	60 Per Cent.	70 Per Cent.	80 Per Cent.	90 Per Cent.	95 Per Cent.	
Five,	848	pound,	34.36	35.56	37.36	38.75	39.85	40.96	42.08	43.54	45.01	47.31	49.35	41.09
		kilogram,	15.59	16.05	16.96	17.58	18.07	18.59	19.10	19.76	20.42	21.47	22.39	18.64
Six,	1,258	pound,	38.01	39.06	41.07	42.64	43.83	45.02	46.27	47.75	49.24	51.75	53.55	45.17
		kilogram,	17.24	17.72	18.63	19.35	19.89	20.43	20.99	21.67	22.34	23.47	24.29	20.49
Seven,	1,419	pound,	40.55	42.54	44.49	46.30	47.60	48.90	50.27	51.97	53.67	56.68	59.00	49.07
		kilogram,	18.40	19.30	20.18	21.01	21.60	22.19	22.80	23.57	24.34	25.71	26.76	22.26
Eight,	1,481	pound,	44.37	46.59	48.95	50.78	52.17	53.57	55.23	57.01	59.25	62.15	64.94	53.92
		kilogram,	20.13	21.14	22.21	23.03	23.67	24.30	25.05	25.86	26.87	28.20	29.46	24.46
Nine,	1,437	pound,	48.89	50.97	53.74	55.55	57.26	58.95	60.62	62.41	64.70	68.06	70.95	59.23
		kilogram,	22.18	23.12	24.38	25.20	25.98	26.74	27.50	28.32	29.35	30.88	32.19	26.87
Ten,	1,563	pound,	52.59	55.07	58.47	60.87	63.10	65.18	67.23	69.24	72.05	75.98	79.13	65.20
		kilogram,	23.86	24.98	26.52	27.61	28.63	29.57	30.50	31.41	32.68	34.47	35.90	29.62
Eleven,	1,293	pound,	56.00	59.45	63.09	65.60	67.74	69.74	71.92	74.16	77.12	81.63	85.95	70.18
		kilogram,	25.67	26.36	28.62	29.79	30.74	31.64	32.63	33.64	34.99	37.04	39.00	31.84
Twelve,	1,253	pound,	61.40	64.29	68.29	71.17	73.40	75.74	78.05	81.24	84.96	91.10	96.46	76.92
		kilogram,	27.85	29.16	30.98	32.29	33.33	34.37	35.41	36.86	38.55	41.33	43.85	34.89
Thirteen,	1,160	pound,	67.23	70.75	74.53	77.47	80.15	82.90	85.97	89.51	94.47	102.39	108.80	84.84
		kilogram,	30.50	32.10	33.81	35.15	36.37	37.62	39.01	40.61	42.85	46.46	49.36	38.49
Fourteen,	908	pound,	72.16	76.38	81.92	85.72	89.47	93.03	96.86	101.74	107.40	116.99	124.83	94.91
		kilogram,	32.74	34.65	37.17	38.90	40.59	42.20	43.94	46.16	48.73	53.07	56.63	42.95
Fifteen,	656	pound,	80.70	85.41	91.78	97.05	101.74	106.00	110.72	115.38	121.76	131.08	139.40	107.10
		kilogram,	36.62	38.76	41.63	44.03	46.16	48.09	50.23	52.34	55.23	59.47	63.24	48.59
Sixteen,	359	pound,	92.95	98.30	108.40	112.89	116.77	121.38	124.51	128.23	132.70	142.03	148.00	121.01
		kilogram,	42.17	44.60	49.18	51.22	52.98	55.06	56.48	58.17	60.66	64.44	67.15	54.90
Seventeen,	192	pound,	101.20	109.90	114.76	118.84	122.38	126.57	131.80	136.77	140.78	150.40	156.40	127.49
		kilogram,	45.91	49.54	52.06	53.92	55.52	57.42	59.79	62.05	63.87	68.23	70.95	57.84
Eighteen,	84	pound,	106.80	115.92	121.87	124.98	128.04	131.71	133.47	138.35	142.01	147.20	157.20	132.55
		kilogram,	48.45	52.60	55.28	56.69	58.09	59.75	61.46	62.77	64.43	66.79	71.32	60.13

TABLE No. 5. — CALCULATED FROM TABLE 8 OF ORIGINAL ARTICLE.

Weights of Boston School-boys of American Parentage.

| Age at Last Birthday. | Number of Observations. | Unit of Weights. | Values at the Following Percentile Grades. | | | | | | | | | | | Average. |
			5 Per Cent.	10 Per Cent.	20 Per Cent.	30 Per Cent.	40 Per Cent.	50 Per Cent.	60 Per Cent.	70 Per Cent.	80 Per Cent.	90 Per Cent.	95 Per Cent.	
Five,	201	pound, kilogram,	34.19 15.51	35.13 15.94	37.00 16.79	38.54 17.48	39.70 18.01	40.87 18.54	42.05 19.08	43.59 19.78	45.14 20.48	47.87 21.72	49.98 22.67	41.20 18.71
Six,	342	pound, kilogram,	38.01 17.24	38.98 17.68	40.94 18.58	42.55 19.31	43.76 19.85	44.97 20.40	46.24 20.98	47.85 21.72	49.46 22.44	52.40 23.77	54.54 24.73	43.14 20.48
Seven,	369	pound, kilogram,	40.37 18.32	42.45 19.26	44.56 20.21	46.46 21.08	47.92 21.75	49.30 22.41	50.91 23.09	52.48 23.81	54.10 24.54	57.24 25.97	59.81 27.13	49.47 22.44
Eight,	407	pound, kilogram,	45.26 20.54	47.00 21.33	49.43 22.43	51.13 23.19	52.61 23.87	54.10 24.54	55.70 25.29	57.42 26.05	59.77 27.11	62.83 28.51	65.15 29.56	54.13 24.70
Nine,	381	pound, kilogram,	48.80 22.14	51.07 23.16	54.11 24.54	55.94 25.38	57.78 26.21	59.41 26.95	61.01 27.67	62.87 28.52	65.11 29.54	69.82 31.67	73.41 33.31	59.97 26.58
Ten,	360	pound, kilogram,	54.00 24.49	56.25 25.51	59.61 27.04	62.64 28.42	64.66 29.28	66.46 30.15	68.31 30.99	70.24 31.87	73.12 33.17	76.86 34.87	80.29 36.43	66.62 30.22
Eleven,	350	pound, kilogram,	58.35 26.47	62.18 28.21	65.36 29.65	67.51 30.63	69.41 31.49	71.52 32.45	73.76 33.46	76.14 34.54	79.45 36.04	84.90 38.52	89.17 40.46	72.39 32.83
Twelve,	373	pound, kilogram,	62.38 28.30	65.16 29.56	69.95 31.73	72.64 32.96	75.32 34.18	78.04 35.41	81.22 36.85	84.34 38.26	88.90 40.34	95.62 43.38	102.24 46.99	79.92 36.21
Thirteen,	391	pound, kilogram,	69.80 31.66	72.94 33.10	76.93 34.90	79.79 36.20	82.45 37.41	85.71 38.89	89.10 40.45	93.83 42.57	99.78 45.26	107.58 48.81	114.23 51.82	88.26 40.04
Fourteen,	386	pound, kilogram,	74.91 33.98	78.74 35.73	84.14 38.17	89.04 40.40	93.22 42.29	97.44 44.21	101.73 46.15	106.83 48.47	114.17 51.80	124.16 56.32	131.85 59.82	99.28 45.03
Fifteen,	342	pound, kilogram,	81.24 36.86	87.20 39.57	95.33 43.26	101.50 46.05	105.84 48.03	110.84 50.29	114.58 51.98	119.57 54.25	126.76 57.50	135.93 61.67	141.96 64.41	110.84 50.26
Sixteen,	232	pound, kilogram,	96.30 43.69	101.28 45.95	110.08 49.95	114.84 52.10	119.29 54.12	123.33 55.95	126.23 57.26	130.76 59.32	136.65 61.99	145.02 65.80	151.00 68.77	123.67 56.09
Seventeen,	128	pound, kilogram,	98.80 44.82	110.36 50.07	115.53 52.42	119.54 54.23	123.29 55.93	127.00 57.89	133.40 60.52	137.57 62.42	140.85 63.91	150.13 68.11	154.00 70.13	128.72 58.40
Eighteen,	65	pound, kilogram,	114.16 51.79	116.80 52.99	122.00 55.35	125.25 56.72	128.00 58.07	131.20 59.52	135.20 61.34	137.80 62.52	142.00 64.43	152.00 68.96	160.80 72.96	132.71 60.20

TABLE No. 6. — CALCULATED FROM TABLE 9 OF ORIGINAL ARTICLE.

Weights of Boston School-boys of Irish Parentage.

AGE AT LAST BIRTHDAY.	Number of Observations.	Unit of Weights.	VALUES AT THE FOLLOWING PERCENTILE GRADES.											Average.
			5 Per Cent.	10 Per Cent.	20 Per Cent.	30 Per Cent.	40 Per Cent.	50 Per Cent.	60 Per Cent.	70 Per Cent.	80 Per Cent.	90 Per Cent.	95 Per Cent.	
Five,	366	pound,	34.64	35.72	37.83	39.04	40.17	41.29	42.51	43.80	45.28	47.70	49.48	41.33
		kilogram,	15.72	16.21	17.17	17.71	18.23	18.73	19.29	19.91	20.55	21.65	22.45	18.75
Six,	503	pound,	38.05	39.16	41.40	42.86	44.03	45.20	46.41	47.69	48.98	51.00	53.33	45.25
		kilogram,	17.26	17.76	18.78	19.45	19.97	20.51	21.06	21.64	22.22	23.13	24.19	20.52
Seven,	562	pound,	41.24	42.77	44.70	46.40	47.61	48.82	50.04	51.71	53.39	56.17	57.87	48.99
		kilogram,	18.71	19.41	20.28	21.05	21.61	22.15	22.70	23.45	24.22	25.48	26.25	22.19
Eight,	588	pound,	45.52	47.08	49.65	51.13	52.46	53.78	55.42	57.14	59.29	61.88	64.73	54.12
		kilogram,	20.51	21.37	22.52	23.19	23.79	24.39	25.14	25.92	26.89	28.07	29.36	24.55
Nine,	556	pound,	48.55	50.79	53.44	55.38	57.13	58.87	60.61	62.45	64.54	67.55	70.04	58.92
		kilogram,	22.03	23.04	24.24	25.12	25.92	26.70	27.50	28.33	29.27	30.65	31.78	26.73
Ten,	571	pound,	52.41	54.83	57.96	60.42	62.78	64.93	67.07	69.12	71.89	75.47	77.95	64.99
		kilogram,	23.78	24.87	26.30	27.41	28.48	29.46	30.43	31.35	32.57	34.24	35.37	29.48
Eleven,	548	pound,	56.37	59.32	62.84	65.08	67.25	69.28	71.55	73.74	76.62	80.54	84.65	69.60
		kilogram,	25.57	26.91	28.51	29.53	30.51	31.47	32.46	33.46	34.76	36.54	38.40	31.56
Twelve,	497	pound,	60.07	63.91	67.90	70.70	72.96	75.18	77.39	80.02	83.33	89.01	93.43	75.70
		kilogram,	27.25	28.99	30.71	32.08	33.10	34.11	35.12	36.31	37.81	40.38	42.39	34.34
Thirteen,	463	pound,	66.66	70.16	73.47	76.19	78.85	81.61	84.65	87.80	91.09	98.70	102.36	82.84
		kilogram,	30.24	31.83	33.33	34.57	35.78	37.03	38.40	39.84	41.59	44.78	46.44	37.58
Fourteen,	334	pound,	71.41	74.88	80.54	84.02	87.03	90.28	93.39	97.25	102.20	108.49	115.53	91.19
		kilogram,	32.40	33.97	36.54	38.12	39.49	40.96	42.37	44.12	46.37	49.22	52.42	41.36
Fifteen,	155	pound,	80.69	84.44	89.00	92.30	95.66	99.60	104.29	108.71	114.00	121.25	125.10	101.21
		kilogram,	36.61	38.31	40.38	41.88	43.39	45.22	47.31	49.32	51.72	55.00	56.76	45.90
Sixteen,	61	pound,	84.00	91.47	97.36	109.07	112.40	115.43	119.28	123.80	128.24	133.12	137.80	112.88
		kilogram,	38.11	41.49	44.17	49.48	51.00	52.38	54.12	56.16	58.18	60.39	62.52	51.19
Seventeen, Eighteen,	26, 5	pound,	102.67	104.90	111.20	115.40	121.20	129.33	132.60	137.40	142.80	145.90	150.00	127.40
		kilogram,	46.50	47.54	50.45	52.27	54.98	58.67	60.16	62.34	64.79	66.20	72.41	57.80

TABLE No. 7. — CALCULATED FROM TABLE 10 OF ORIGINAL ARTICLE.

Heights of Boston School-girls irrespective of Nationality.

Age at Last Birthday	Number of Observations	Unit of Measurement	Values at the Following Percentile Grades.											Average.
			5 Per Cent.	10 Per Cent.	20 Per Cent.	30 Per Cent.	40 Per Cent.	50 Per Cent.	60 Per Cent.	70 Per Cent.	80 Per Cent.	90 Per Cent.	95 Per Cent.	
Five,	605	inch,	38.27	39.08	39.90	40.45	40.95	41.43	41.90	42.41	42.04	43.74	44.58	41.29
		c. m.,	97.2	99.3	101.3	102.7	104.0	105.2	106.4	107.7	110.1	111.1	113.2	104.9
Six,	987	inch,	40.30	41.10	41.91	42.45	42.05	43.40	43.85	44.41	45.03	45.90	46.71	43.35
		c. m.,	102.4	104.4	106.5	107.8	109.1	110.2	111.4	112.8	114.4	116.6	118.6	110.1
Seven,	1,199	inch,	42.11	43.12	43.95	44.70	45.14	45.63	46.14	46.73	47.37	48.14	49.05	45.52
		c. m.,	107.7	109.5	111.6	113.5	114.7	115.9	117.2	118.7	120.3	122.3	124.6	115.6
Eight,	1,299	inch,	44.08	44.91	45.94	46.53	47.09	47.66	48.23	48.84	49.50	50.56	51.37	47.58
		c. m.,	112.0	114.1	116.7	118.2	119.6	121.0	122.5	124.1	125.9	128.4	130.5	120.9
Nine,	1,140	inch,	45.81	46.56	47.67	48.41	49.01	49.64	50.08	50.65	51.32	52.33	53.15	49.37
		c. m.,	116.4	118.3	121.1	123.0	124.5	125.8	127.2	128.0	130.4	132.9	135.0	125.4
Ten,	1,089	inch,	47.49	48.37	49.49	50.26	50.88	51.46	52.04	52.65	53.41	54.58	55.60	51.34
		c. m.,	120.6	122.9	125.7	127.7	129.2	130.7	132.2	133.7	135.7	138.6	141.4	130.4
Eleven,	936	inch,	49.33	50.27	51.35	52.14	52.73	53.41	54.16	54.89	55.76	57.05	57.90	53.42
		c. m.,	125.3	127.7	130.4	132.4	133.9	135.7	137.6	139.4	141.6	144.9	147.2	135.7
Twelve,	935	inch,	51.25	52.24	53.45	54.33	55.15	55.88	56.59	57.28	58.39	59.73	60.81	55.88
		c. m.,	130.2	132.7	135.8	138.0	140.1	141.9	143.7	145.7	148.3	151.7	154.5	141.9
Thirteen,	830	inch,	53.61	54.50	55.75	56.80	57.61	58.40	59.19	59.91	60.75	61.79	62.80	58.16
		c. m.,	136.2	138.4	141.6	144.3	146.3	148.3	150.3	152.2	154.3	156.9	159.5	147.7
Fourteen,	675	inch,	55.87	56.98	58.11	58.90	59.59	60.20	60.71	61.28	61.04	62.00	64.05	50.94
		c. m.,	141.9	144.7	147.6	149.6	151.4	152.9	154.2	155.7	157.3	160.0	162.7	152.3
Fifteen,	450	inch,	57.39	58.29	59.33	60.10	60.67	61.21	61.73	62.32	63.15	64.15	65.00	61.10
		c. m.,	145.8	148.1	150.7	152.7	154.1	155.5	156.8	158.3	160.4	162.9	165.1	155.2
Sixteen,	353	inch,	57.82	58.04	59.72	60.49	61.16	61.78	62.37	62.95	63.68	64.77	65.59	61.59
		c. m.,	146.9	148.9	151.7	153.6	155.3	156.9	158.4	159.9	161.7	164.5	166.6	156.4
Seventeen,	233	inch,	58.22	59.08	59.05	60.88	61.58	62.18	62.68	63.21	63.83	64.88	65.85	61.92
		c. m.,	147.9	150.1	152.3	154.6	156.4	157.9	159.2	160.6	162.1	164.8	167.3	157.2
Eighteen,	155	inch,	58.67	59.31	60.12	60.77	61.33	61.85	62.39	62.95	64.14	64.88	65.87	61.95
		c. m.,	149.0	150.6	152.7	154.4	155.8	157.1	158.5	159.9	162.9	164.8	167.3	157.3

TABLE No. 8. — CALCULATED FROM TABLE 11 OF ORIGINAL ARTICLE.

Heights of Boston School-girls of American Parentage.

| Age at Last Birthday. | Number of Observations. | Unit of Measurement. | Values at the following percentile grades. | | | | | | | | | | | Average. |
			5 Per Cent.	10 Per Cent.	20 Per Cent.	30 Per Cent.	40 Per Cent.	50 Per Cent.	60 Per Cent.	70 Per Cent.	80 Per Cent.	90 Per Cent.	95 Per Cent.	
Five,	127	inch,	38.55	39.28	40.14	40.67	41.17	41.60	42.06	42.66	43.31	44.04	44.94	41.47
		c.m.,	97.9	99.8	102.0	103.3	104.6	105.7	106.8	108.4	110.3	111.9	114.2	105.3
Six,	236	inch,	40.57	41.27	42.13	42.62	43.11	43.50	44.10	44.75	45.13	46.39	47.36	43.60
		c.m.,	103.1	104.8	107.0	108.3	109.5	110.7	112.0	113.7	115.4	117.8	120.3	110.9
Seven,	346	inch,	42.71	43.43	44.35	45.01	45.52	46.05	46.57	47.12	47.78	48.71	49.51	45.94
		c.m.,	108.5	110.3	112.7	114.3	115.7	117.0	118.3	119.7	121.4	123.7	125.8	116.7
Eight,	358	inch,	44.33	45.25	46.29	47.01	47.60	48.21	48.96	49.50	50.18	51.05	51.92	48.07
		c.m.,	112.6	114.9	117.6	119.4	120.9	122.4	124.1	125.7	127.5	129.7	131.6	122.1
Nine,	323	inch,	46.21	46.89	48.08	48.61	49.16	49.70	50.27	50.90	51.64	52.81	53.69	49.61
		c.m.,	117.4	119.1	122.1	123.5	124.9	126.2	127.7	129.2	131.2	134.1	136.4	126.0
Ten,	336	inch,	47.91	48.82	49.87	50.70	51.32	51.90	52.50	53.14	53.93	54.98	56.32	51.78
		c.m.,	121.7	124.0	126.7	128.8	130.3	131.8	133.3	135.0	137.0	139.6	143.0	131.5
Eleven,	290	inch,	49.00	50.48	51.52	52.37	53.11	53.92	54.63	55.42	56.36	57.52	58.37	53.79
		c.m.,	126.2	128.2	130.9	133.0	134.9	137.0	138.8	140.8	143.1	146.1	148.3	136.6
Twelve,	309	inch,	52.18	53.15	54.10	55.12	55.91	56.61	57.34	58.20	59.31	60.51	61.62	57.16
		c.m.,	132.5	135.0	137.4	140.0	142.0	143.8	145.6	147.0	150.6	153.7	156.5	145.2
Thirteen,	307	inch,	54.07	54.88	56.11	57.44	58.25	59.10	59.75	60.46	61.27	62.41	63.35	58.75
		c.m.,	137.3	139.4	143.3	145.9	147.9	150.1	151.8	153.6	155.6	158.5	160.9	149.2
Fourteen,	290	inch,	56.69	57.71	58.69	59.40	60.03	60.47	60.91	61.54	62.26	63.27	64.57	60.32
		c.m.,	144.0	146.6	149.1	150.9	152.5	153.6	154.7	156.3	158.1	160.7	163.2	153.2
Fifteen,	255	inch,	58.11	58.91	59.75	60.39	60.95	61.43	61.89	62.53	63.30	64.36	65.17	61.30
		c.m.,	147.6	149.6	151.8	153.4	154.8	156.0	157.2	158.8	160.8	163.5	165.5	155.9
Sixteen,	238	inch,	57.99	58.66	59.70	60.55	61.31	62.02	62.45	63.08	63.50	64.86	65.65	61.72
		c.m.,	147.3	149.0	151.6	153.8	155.7	157.5	158.6	160.2	162.0	164.7	166.7	156.7
Seventeen,	168	inch,	58.11	59.11	60.14	61.06	61.76	62.32	62.82	63.34	63.89	64.89	65.82	61.99
		c.m.,	147.7	150.1	152.8	155.1	156.9	158.3	159.6	160.9	162.3	164.8	167.2	157.4
Eighteen,	118	inch,	58.58	59.23	60.33	61.02	61.47	61.92	62.47	63.08	64.24	64.90	65.08	62.01
		c.m.,	148.8	150.8	153.2	155.0	156.1	157.3	158.7	160.2	163.2	164.9	166.8	157.5

TABLE No. 9.—CALCULATED FROM TABLE 12 OF ORIGINAL ARTICLE.

Heights of Boston School-girls of Irish Parentage.

Age at Last Birthday.	Number of Observations.	Unit of Measurement.	Values at the Following Percentile Grades.											Average.
			5 Per Cent.	10 Per Cent.	20 Per Cent.	30 Per Cent.	40 Per Cent.	50 Per Cent.	60 Per Cent.	70 Per Cent.	80 Per Cent.	90 Per Cent.	95 Per Cent.	
Five,	236	inch,	38.21	39.07	40.06	40.49	40.92	41.34	41.76	42.20	42.68	43.32	43.83	41.18
		c. m.,	97.0	99.2	101.7	102.8	103.9	105.0	106.1	107.2	108.4	110.0	111.3	104.6
Six,	395	inch,	40.23	41.08	41.76	42.36	42.92	43.39	43.85	44.12	45.06	45.80	46.48	43.29
		c. m.,	102.2	104.3	106.1	107.6	109.0	110.2	111.4	112.8	114.4	116.3	118.1	109.9
Seven,	426	inch,	42.51	43.23	44.00	44.54	45.07	45.52	45.97	46.63	47.29	47.94	48.99	46.45
		c. m.,	108.0	109.8	111.8	113.1	114.5	115.6	116.8	118.4	120.1	121.8	124.4	115.4
Eight,	486	inch,	43.86	44.65	45.71	46.37	46.90	47.46	48.03	48.65	49.36	50.31	50.98	47.39
		c. m.,	111.4	113.4	116.1	117.8	119.1	120.5	122.0	123.6	125.4	127.8	129.4	120.4
Nine,	416	inch,	45.58	46.43	47.63	48.36	49.10	49.60	50.09	50.69	51.15	52.16	52.90	49.27
		c. m.,	115.8	117.9	120.7	122.8	124.7	126.0	127.2	128.5	129.9	132.5	134.4	125.2
Ten,	379	inch,	47.49	48.52	49.08	50.34	50.86	51.37	51.87	52.42	52.97	54.21	55.00	51.20
		c. m.,	120.6	123.2	126.2	127.9	129.2	130.5	131.7	133.1	134.5	137.7	139.7	130.1
Eleven,	340	inch,	49.50	50.77	51.35	52.03	52.55	53.08	53.73	54.38	55.03	56.18	57.47	53.13
		c. m.,	125.7	127.9	130.4	132.2	133.5	134.8	136.5	138.1	139.8	142.7	146.0	134.9
Twelve,	297	inch,	51.12	51.93	53.19	53.98	54.76	55.49	56.18	56.92	57.84	59.31	60.47	55.41
		c. m.,	129.8	131.9	135.1	137.1	139.1	140.9	142.7	144.6	146.9	150.7	153.6	140.8
Thirteen,	278	inch,	53.39	54.37	55.49	56.33	57.05	57.75	58.49	59.30	60.18	61.19	62.02	57.64
		c. m.,	135.6	138.1	140.9	143.1	144.9	146.7	148.6	150.6	152.9	155.4	157.5	146.3
Fourteen,	192	inch,	55.60	56.80	57.72	58.47	59.21	60.03	60.56	61.11	61.76	62.74	63.54	59.67
		c. m.,	141.4	144.3	146.6	148.5	150.4	152.5	153.8	155.2	156.8	159.4	161.4	151.5
Fifteen,	95	inch,	56.23	57.50	58.70	59.50	60.16	60.66	61.15	61.62	62.29	63.64	64.73	60.47
		c. m.,	142.8	146.0	149.1	151.1	152.8	154.1	155.3	156.5	158.2	161.6	164.4	153.5
Sixteen,	49	inch,	57.29	58.90	60.06	60.47	60.88	61.22	61.76	62.26	62.80	64.03	64.82	61.05
		c. m.,	145.3	149.6	152.5	153.6	154.6	155.7	156.9	158.1	159.5	162.6	164.7	155.1
Seventeen,	18	inch,	58.20	59.10	59.70	60.60	61.32	61.80	62.35	62.95	64.10	65.00	66.80	62.00
Eighteen,	6	c. m.,	147.8	150.1	151.6	153.9	155.7	157.0	158.4	159.9	162.8	166.6	169.7	157.5

TABLE No. 10.—CALCULATED FROM TABLE 13 OF ORIGINAL ARTICLE.

Weights of Boston School-girls irrespective of Nationality.

AGE AT LAST BIRTHDAY.	Number of Observations.	Unit of Weights.	Values at the following Percentile Grades.											Average.
			5 Per Cent.	10 Per Cent.	20 Per Cent.	30 Per Cent.	40 Per Cent.	50 Per Cent.	60 Per Cent.	70 Per Cent.	80 Per Cent.	90 Per Cent.	95 Per Cent.	
Five,	605	pound,	32.45	34.31	35.82	37.34	38.59	39.63	40.67	41.72	43.44	45.43	47.88	39.66
		kilogram,	14.72	15.56	16.25	16.94	17.51	17.98	18.45	18.93	19.71	20.62	21.73	17.99
Six,	987	pound,	35.42	37.13	39.04	40.43	41.83	43.11	44.37	45.63	47.56	49.78	52.36	43.28
		kilogram,	16.07	16.85	17.71	18.22	18.98	19.56	20.13	20.71	21.58	22.58	23.75	19.63
Seven,	1,199	pound,	39.04	40.04	42.87	44.37	45.88	47.30	48.71	50.19	52.45	55.36	57.56	47.46
		kilogram,	17.71	18.44	19.45	20.13	20.02	21.47	22.10	22.77	23.79	25.11	26.11	27.10
Eight,	1,299	pound,	42.61	44.25	46.85	48.33	50.31	51.45	53.45	55.27	57.24	60.32	62.60	52.04
		kilogram,	19.34	20.07	21.26	21.93	22.82	23.33	24.24	25.07	25.97	27.37	28.40	23.44
Nine,	1,149	pound,	46.16	48.21	51.16	53.22	55.00	56.02	58.28	60.22	62.57	65.83	69.96	57.07
		kilogram,	20.94	21.88	23.20	24.14	24.95	25.68	26.44	27.37	28.39	29.87	31.74	25.91
Ten,	1,089	pound,	50.26	52.19	55.26	57.64	59.56	61.40	62.70	66.24	69.03	73.43	76.89	62.35
		kilogram,	22.80	23.68	25.07	26.15	27.01	27.85	28.90	30.05	31.31	33.32	34.88	28.29
Eleven,	936	pound,	54.08	56.16	59.61	62.42	64.92	67.02	70.48	73.36	77.07	83.71	89.27	68.84
		kilogram,	23.90	25.47	27.04	28.32	29.45	30.62	31.98	33.28	34.97	37.98	40.50	31.23
Twelve,	935	pound,	59.60	62.97	67.32	70.64	73.53	77.08	80.50	84.04	88.23	94.71	102.73	78.31
		kilogram,	27.07	28.57	30.55	32.05	33.36	34.98	36.53	38.13	40.03	41.21	46.61	35.53
Thirteen,	820	pound,	65.75	70.37	75.16	79.33	83.92	88.02	91.06	96.46	102.12	108.47	115.42	88.65
		kilogram,	29.83	31.93	34.10	35.99	38.08	39.94	41.72	43.76	46.33	49.21	52.37	40.21
Fourteen,	675	pound,	76.57	80.36	85.76	90.14	94.00	97.75	101.23	105.08	109.94	117.12	124.73	98.43
		kilogram,	34.74	36.46	38.91	40.89	42.64	44.35	45.92	47.95	49.88	53.14	56.58	44.65
Fifteen,	450	pound,	83.39	88.80	95.43	99.59	102.49	105.11	108.43	112.39	117.64	124.14	132.22	106.08
		kilogram,	37.84	40.29	43.30	45.18	46.50	47.69	49.20	51.00	53.38	56.31	59.99	48.12
Sixteen,	353	pound,	88.47	93.51	99.05	103.50	107.28	111.27	115.79	119.95	124.79	130.99	135.94	112.03
		kilogram,	40.14	42.42	44.93	46.96	48.68	50.48	52.54	54.42	56.61	59.43	61.68	50.81
Seventeen,	233	pound,	93.20	97.75	104.32	107.57	110.02	112.76	117.20	122.75	127.54	135.88	141.15	115.53
		kilogram,	42.28	44.35	47.33	48.81	49.92	51.16	53.18	55.69	57.86	61.65	64.04	52.41
Eighteen,	155	pound,	95.29	98.40	102.36	105.18	110.16	112.64	116.80	122.88	127.33	133.45	138.80	115.16
		kilogram,	43.19	44.64	46.44	47.72	49.98	51.11	52.99	55.75	57.77	60.54	62.97	52.24

TABLE No. 11. — CALCULATED FROM TABLE 14 OF ORIGINAL ARTICLE.

Weights of Boston School-girls of American Parentage.

Age at Last Birthday.	Number of Observations.	Unit of Weights.	Values at the following Percentile Grades.											Average.
			5 Per Cent.	10 Per Cent.	20 Per Cent.	30 Per Cent.	40 Per Cent.	50 Per Cent.	60 Per Cent.	70 Per Cent.	80 Per Cent.	90 Per Cent.	95 Per Cent.	
Five,	127	pound,	32.52	34.34	35.92	37.51	38.64	39.56	40.49	41.41	42.97	45.64	50.40	39.82
		kilogram,	14.76	15.57	16.30	17.02	17.53	17.94	18.37	18.79	19.50	20.71	22.86	18.06
Six,	236	pound,	35.60	37.37	39.22	40.72	42.18	43.14	44.70	45.96	48.31	51.48	53.84	43.81
		kilogram,	16.19	16.96	17.79	18.48	19.14	19.71	20.28	20.86	21.93	23.35	24.42	19.87
Seven,	346	pound,	39.43	41.16	43.25	44.90	46.43	47.74	49.04	50.01	52.92	56.23	58.72	48.02
		kilogram,	17.89	18.67	19.62	20.37	21.07	21.67	22.25	22.96	24.01	25.50	26.84	21.78
Eight,	338	pound,	42.85	44.68	47.27	49.26	50.99	52.56	54.15	56.01	57.36	61.47	65.40	52.93
		kilogram,	19.45	20.27	21.45	22.35	23.13	23.84	24.56	25.40	26.25	27.88	29.95	24.01
Nine,	323	pound,	46.60	49.51	51.14	53.01	54.81	56.53	58.38	60.97	63.65	67.34	72.27	57.52
		kilogram,	21.14	22.01	23.19	24.05	24.86	25.74	26.48	27.66	28.87	30.55	32.79	26.10
Ten,	336	pound,	50.36	52.33	55.68	58.18	60.39	62.65	65.10	67.84	71.53	77.37	86.60	64.09
		kilogram,	22.84	23.88	25.26	26.39	27.40	28.42	29.54	30.78	32.45	35.11	39.29	29.07
Eleven,	290	pound,	54.25	56.67	60.05	63.06	65.43	68.75	72.11	75.41	79.40	86.73	92.75	70.26
		kilogram,	24.60	25.70	27.24	28.61	29.69	31.19	32.72	34.22	36.02	39.35	42.08	31.87
Twelve,	309	pound,	62.84	65.79	69.61	72.77	76.48	79.90	83.07	86.62	90.40	97.70	106.10	81.35
		kilogram,	28.51	29.85	31.58	33.02	34.70	36.25	37.69	39.30	41.01	44.33	48.14	36.90
Thirteen,	307	pound,	66.43	71.04	77.66	81.88	86.75	90.88	95.60	100.53	104.28	110.92	116.53	91.18
		kilogram,	30.14	32.23	35.24	37.15	39.36	41.23	43.37	45.61	47.35	50.33	52.87	41.36
Fourteen,	290	pound,	79.29	83.05	88.07	92.14	96.21	99.88	103.52	107.22	110.94	117.76	125.00	100.32
		kilogram,	35.97	37.68	39.95	41.80	43.65	45.31	46.97	48.65	50.34	53.43	56.71	45.50
Fifteen,	255	pound,	86.87	92.14	98.48	102.30	104.52	107.13	110.67	115.06	119.00	125.82	132.80	108.42
		kilogram,	39.41	41.80	44.68	46.42	47.42	48.61	50.21	52.20	54.26	57.08	60.26	49.17
Sixteen,	238	pound,	88.15	93.40	100.03	104.62	108.43	112.80	116.97	120.93	125.66	131.28	136.26	112.97
		kilogram,	40.00	42.37	45.38	47.46	49.20	51.16	53.07	54.86	57.01	59.56	61.82	51.24
Seventeen,	168	pound,	93.52	97.90	104.53	108.16	110.99	113.68	118.70	122.96	127.96	135.83	141.00	115.84
		kilogram,	42.43	44.42	47.42	49.07	50.36	51.58	53.86	55.79	58.06	61.63	64.00	52.54
Eighteen,	118	pound,	95.93	99.02	102.85	105.62	110.71	113.33	117.91	123.72	128.20	133.64	138.40	115.80
		kilogram,	43.52	44.92	46.67	47.92	50.23	51.42	53.50	56.13	58.16	60.63	62.79	52.52

TABLE No. 12. — CALCULATED FROM TABLE 15 OF ORIGINAL ARTICLE.

Weights of Boston School-girls of Irish Parentage.

AGE AT LAST BIRTHDAY.	Number of Observations.	Unit of Weights.	VALUES AT THE FOLLOWING PERCENTILE GRADES.											Average.
			5 Per Cent.	10 Per Cent.	20 Per Cent.	30 Per Cent.	40 Per Cent.	50 Per Cent.	60 Per Cent.	70 Per Cent.	80 Per Cent.	90 Per Cent.	95 Per Cent.	
Five,	236	pound,	32.78	34.43	35.96	37.47	38.64	39.61	40.58	41.55	43.14	45.24	47.16	39.63
		kilogram,	14.87	15.62	16.31	17.00	17.53	17.97	18.41	18.85	19.57	20.53	21.40	17.97
Six,	395	pound,	35.48	37.24	39.11	40.52	41.93	43.18	44.42	45.67	47.47	49.47	51.60	43.21
		kilogram,	16.10	16.90	17.74	18.39	19.02	19.59	20.15	20.72	21.54	22.44	23.40	19.60
Seven,	426	pound,	39.50	41.36	43.12	44.47	45.82	47.20	48.77	50.59	52.73	55.60	57.41	47.64
		kilogram,	17.92	18.76	19.56	20.17	20.79	21.42	22.13	22.86	23.92	25.26	26.10	21.61
Eight,	486	pound,	42.31	43.74	46.43	48.48	50.39	51.94	53.48	55.30	57.26	60.25	62.14	51.80
		kilogram,	19.20	19.85	21.07	22.00	22.86	23.56	24.26	25.09	25.98	27.33	28.19	23.50
Nine,	416	pound,	45.96	48.17	51.20	53.25	55.01	56.59	58.20	60.05	61.90	65.43	68.99	56.76
		kilogram,	20.86	21.86	23.22	24.15	24.95	25.67	26.77	27.24	28.08	29.69	31.30	25.75
Ten,	379	pound,	50.40	52.35	55.45	57.98	59.67	61.36	63.51	65.95	68.29	71.35	73.87	61.59
		kilogram,	22.86	23.75	25.15	26.30	27.06	27.83	28.81	29.92	30.98	32.37	33.51	27.94
Eleven,	340	pound,	54.25	56.57	59.56	62.13	64.36	66.68	69.25	72.18	75.76	80.91	85.06	67.83
		kilogram,	24.60	25.57	27.02	28.19	29.19	30.25	31.41	32.75	34.37	36.71	38.87	30.77
Twelve,	307	pound,	50.17	61.63	66.04	69.03	71.98	75.09	78.63	82.33	86.95	92.29	97.85	76.15
		kilogram,	26.84	27.96	29.96	31.31	32.65	34.07	35.68	37.36	39.00	41.87	44.40	34.55
Thirteen,	278	pound,	64.18	68.80	73.93	77.22	81.53	85.86	89.83	93.32	97.45	103.89	109.60	85.76
		kilogram,	29.11	31.21	33.54	35.04	36.99	38.96	40.76	42.34	44.21	47.13	49.72	38.91
Fourteen,	192	pound,	74.80	79.25	83.05	88.40	92.33	95.90	99.56	103.60	108.26	113.93	122.53	96.36
		kilogram,	33.93	35.95	37.68	40.11	41.89	43.51	45.16	47.00	49.12	51.69	55.59	43.71
Fifteen,	95	pound,	77.40	83.25	88.67	93.33	97.56	100.62	103.33	105.87	110.50	117.33	130.40	100.46
		kilogram,	35.12	37.77	40.23	42.34	44.25	45.65	46.88	48.03	50.14	53.24	59.16	45.56
Sixteen,	49	pound,	87.60	93.80	97.44	99.98	101.84	105.60	109.62	115.73	120.56	130.20	136.00	108.56
		kilogram,	39.75	42.55	44.21	45.31	46.20	47.91	49.69	52.51	54.69	59.07	61.70	49.24
Seventeen, Eighteen,	18 6	pound,	98.40	100.80	104.40	106.80	108.40	110.00	111.60	113.20	124.40	141.40	161.20	115.82
		kilogram,	44.64	45.73	47.36	48.45	49.18	49.91	50.63	51.36	56.43	65.51	73.14	52.49

A geometrical construction of a special case will perhaps best serve to place the matter in a clear light. Let us suppose one thousand grown men standing in line arranged according to height. The heads of these men will form a curved line represented in its general form by the curve ST in Fig 1. In this diagram the line SO represents the height

Fig. 1.

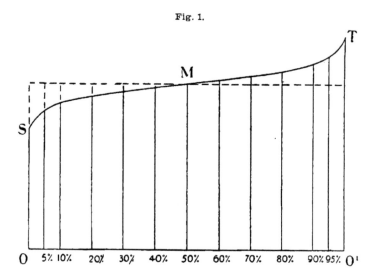

of the shortest and the line TO¹ that of the tallest man. The curve ST, representing the heights of the intermediate men, is approximately a straight line in a large part of its course but bends up sharply at the right and down sharply at the left owing to the presence of a few very tall and a few very short men. Mediocrity is the rule and extremes the exception in height as in everything else.

If now we divide this row of men into two equal parts and ascertain the height of the five hundredth man in the row (or, more accurately speaking, the height half way between that of the five hundredth and that of the five hundred and first man) we shall have a value below which one half and above which the other half of the observations lie. This value is termed by Galton the value of the fifty percentile grade, or the median value, and is designated by the letter M. In the same way the values at other percentile grades may be determined by dividing the row at points corresponding to various percentages of the total number of observations. The percentile grades indicated in Fig. 1 are those adopted by Galton, and are practically sufficient to indicate the character of the curve. With a very large number of observations it would of course be possible to determine values below five per cent. and above ninety-five per cent., but in anthropometrical investigations with existing data it does not seem wise to go beyond these limits.

It is evident that the value M will tend to approximate to the average value of all the observations and will be identical with it when the curve ST is symmetrically disposed on both sides of M, *i. e.*, when the values at sixty, seventy, eighty, ninety and ninety-five per cent. exceed M by the same amount, respectively, by which the values at forty, thirty, twenty, ten and five per cent. fall short of it. If A represent the average value of all the observations, then the value of M—A will be a measure of the direction and extent of the asymmetry of the curve ST, *for this value will be zero when the curve is symmetrical, positive when the values at the lower percentile grades fall short of M more than those at the higher grades exceed it, and negative when the reverse is the case.*

Let us now apply this test to the data in our possession, confining our attention for the present to tables 1, 4, 7 and 10, which give the total number of observations irrespective of nationality. By subtracting the average from the median values in these four tables the following table (No. 12a) has been constructed : —

TABLE 12a. — *Values of M—A.*

AGE AT LAST BIRTHDAY.	HEIGHTS IN INCHES.		WEIGHTS IN POUNDS.	
	Boys.	Girls.	Boys.	Girls.
Five, ...	+0.10	+0.14	—0.13	—0.03
Six, ...	+0.12	+0.05	—0.15	—0.17
Seven, ...	+0.09	+0.11	—0.17	—0.16
Eight, ...	+0.08	+0.07	—0.35	—0.59
Nine, ...	+0.08	+0.17	—0.28	—0.45
Ten, ...	+0.05	+0.12	—0.12	—0.95
Eleven, ...	+0.07	—0.01	—0.44	—1.22
Twelve, ...	0.00	0.00	—1.18	—1.23
Thirteen, ...	—0.23	+0.24	—1.94	—0.63
Fourteen, ...	—0.28	+0.26	—1.88	—0.68
Fifteen, ...	+0.07	+0.11	—1.10	—0.97
Sixteen, ...	+0.35	+0.19	+0.37	—0.76
Seventeen, . . .	+0.05	+0.26	—0.92	—2.77
Eighteen, ...	+0.13	—0.10	—0.84	—2.52

An examination of this table or of the curves constructed
from it, as given in Plate 1, shows that the asymmetry of

PLATE I.

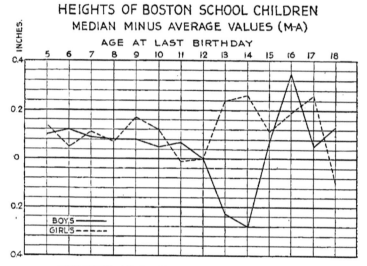

HEIGHTS OF BOSTON SCHOOL CHILDREN
MEDIAN MINUS AVERAGE VALUES (M-A)

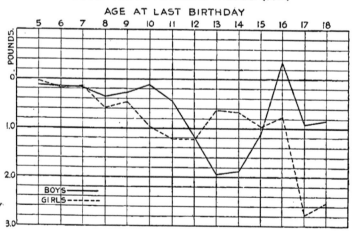

WEIGHTS OF BOSTON SCHOOL CHILDREN.
MEDIAN MINUS AVERAGE VALUES (M-A)

the curves of percentile grades varies very much, at different ages, both in direction and amount. The variation in the value of M—A in the curves of height is much the same as that in the curves of weight for each sex considered by itself, but there is a great difference between the two sexes. This difference shows itself most distinctly between the ages of eleven and fifteen years. During this time a rise in the curves for the males coincides with a fall in those for the females, while before and after this period the curves, as a rule, rise and fall together. We must conclude, therefore, that the rate of annual increase both in height and weight is different at different percentile grades, or, in other words, that large children grow differently from small ones, and moreover, that between the ages of eleven and fifteen years there is a striking difference in the mode of growth of the two sexes. The significance of this conclusion will be made clearer by an examination of the curves constructed directly from the tables of percentile grades. Curves of this sort are presented in plates 2, 3, 4 and 5, constructed from tables 1, 4, 7 and 10, containing the total number of observations irrespective of nationality. Similar curves have been obtained from the remaining tables in which the observations are grouped according to the nationality of the parents, but as they are less regular, owing to the smaller number of observations from which they are constructed, and lead to no additional conclusions, it has not been thought worth while to present them.

A glance at the curves on plates 2–5 shows at once the nature of the asymmetry, the existence of which is indicated by the curves on Plate 1. It will be observed that during the earlier years of school life the curves for the successive years are fairly symmetrical, which is in harmony with the previous observation that in these years the value of M—A does not differ widely from zero. At about ten years of age in girls and eleven or twelve years in boys, the curves become distinctly asymmetrical, owing to the values increasing more rapidly at the higher than at the lower percentile grades. At the age of twelve or thirteen years in girls and fourteen or fifteen years in boys an asymmetry in the opposite direction shows itself, since at this period the values are increasing

PLATE 2. (FROM TABLE I.)

HEIGHTS OF BOSTON SCHOOLBOYS.
IRRESPECTIVE OF NATIONALITY.

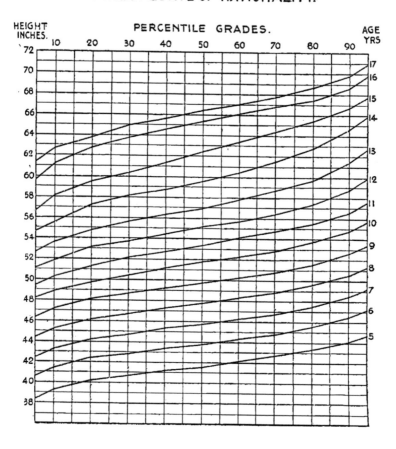

PLATE 3.(FROM TABLE 4.)

WEIGHTS OF BOSTON SCHOOLBOYS.
IRRESPECTIVE OF NATIONALITY.

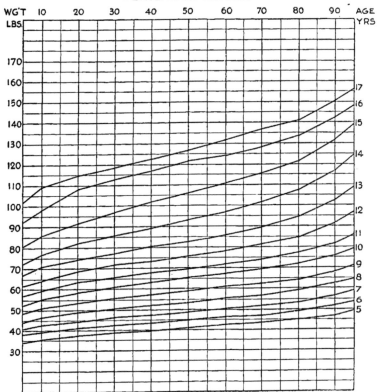

PERCENTILE GRADES.

PLATE 4. (FROM TABLE 7)

HEIGHTS OF BOSTON SCHOOLGIRLS.
IRRESPECTIVE OF NATIONALITY.

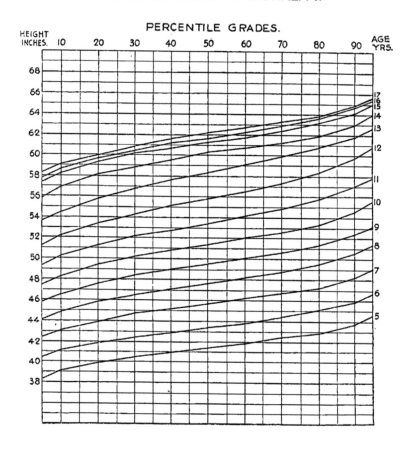

PLATE 5. (FROM TABLE 10.)

WEIGHTS OF BOSTON SCHOOLGIRLS:
IRRESPECTIVE OF NATIONALITY.

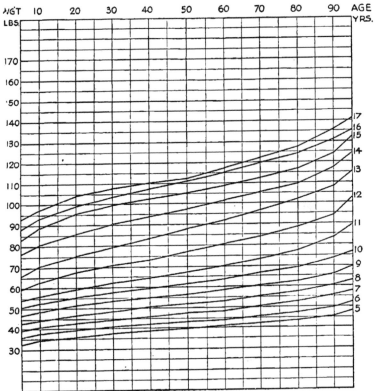

more rapidly at the lower than at the higher percentile grades. These changes correspond accurately with the fall and rise in the value of M—A, as shown on Plate 1.

In the original article on the growth of children it was shown that about two years before the age of puberty there is a period during which the growth in both height and weight shows a distinct acceleration. Now, the rate of growth at the various percentile grades is represented on plates 2–5 by the vertical distances between the curves corresponding to the successive years ; and an inspection of these curves shows that the prepubertal period of accelerated growth, already shown to exist by a comparison of average heights and weights at different ages, occurs all along the line, but that it occurs earlier at the higher than at the lower percentile grades. In other words, we find that the above-mentioned variations in the value of M—A are due to the fact that *the period of acceleration, which is such a distinct phenomenon in the growth of children, occurs at an earlier age in large than in small children.*

The significance of this observation will be best understood by an examination of the annual increase in height and weight of children at the different percentile grades. These values, which are readily obtained from the preceding tables by subtracting the height or weight at any year from that of the year next following, are shown in tables 13, 14, 15 and 16, which have been calculated from tables 1, 4, 7 and 10 respectively.*

---

* It will be noticed that in tables 1, 4, 7 and 10 the ages are given " at the last birth-day." Hence the average age of the children thus grouped together will be six months greater than the age given in the tables. For instance, in Table 1, five years six months is the average age of the 848 boys whose heights at various percentile grades are given in the first line. Now since the figures in the tables of annual increase are the differences between the successive heights and weights in tables 1, 4, 7 and 10, it is evident that they express the yearly growth precisely at the age given in the tables. The first line in Table 13, for instance, is the annual growth in height of boys of six years of age.

TABLE 13.— *Showing Annual Growth in Inches of Boston School-boys irrespective of Nationality.*

AGE.	VALUES AT THE FOLLOWING PERCENTILE GRADES.											Average.
	5 Per Cent.	10 Per Cent.	20 Per Cent.	30 Per Cent.	40 Per Cent.	50 Per Cent.	60 Per Cent.	70 Per Cent.	80 Per Cent.	90 Per Cent.	95 Per Cent.	
Six,	2.27	2.22	2.16	2.17	2.16	2.20	2.22	2.14	2.18	2.34	2.30	2.18
Seven,	1.82	1.84	1.89	1.87	1.92	1.96	1.99	2.02	2.01	1.99	2.14	1.99
Eight,	1.98	1.97	1.95	2.05	2.04	2.01	2.00	2.02	2.06	2.10	2.15	2.02
Nine,	1.95	1.94	1.89	1.89	1.92	1.93	1.94	1.97	1.90	2.03	2.04	1.93
Ten,	1.70	1.80	1.73	1.75	1.83	1.96	2.04	2.07	2.16	2.24	2.37	1.99
Eleven,	1.36	1.34	1.53	1.63	1.65	1.67	1.70	1.79	1.74	1.69	1.67	1.65
Twelve,	1.63	1.64	1.77	1.70	1.71	1.71	1.71	1.73	1.86	2.21	2.38	1.78
Thirteen,	1.52	1.62	1.74	1.91	1.91	1.87	2.09	2.29	2.44	2.72	2.85	2.10
Fourteen,	1.95	2.08	2.32	2.38	2.41	2.02	2.04	2.76	3.00	3.14	3.16	2.67
Fifteen,	1.98	2.41	2.33	2.32	2.55	2.77	2.87	2.90	2.68	2.24	2.01	2.42
Sixteen,	3.21	3.14	3.34	3.36	3.25	2.98	2.71	2.33	1.99	1.80	1.96	2.70
Seventeen,	1.64	1.39	1.05	1.23	0.99	0.86	0.95	1.23	1.27	1.33	1.08	1.16

TABLE 14.— *Showing Annual Increase in Pounds of Boston School-boys irrespective of Nationality.*

Age	Values at the Following Percentile Grades.											Average.
	5 Per Cent.	10 Per Cent.	20 Per Cent.	30 Per Cent.	40 Per Cent.	50 Per Cent.	60 Per Cent.	70 Per Cent.	80 Per Cent.	90 Per Cent.	95 Per Cent.	
Six,	3.65	3.70	3.71	3.89	3.98	4.06	4.19	4.21	4.23	4.44	4.20	4.08
Seven,	2.54	3.48	3.42	3.66	3.77	3.88	4.00	4.22	4.43	4.93	5.45	3.90
Eight,	3.82	4.05	4.46	4.43	4.57	4.67	4.96	5.04	5.58	5.47	5.94	4.85
Nine,	4.52	4.38	4.79	4.77	5.09	5.38	5.39	5.40	5.45	5.91	6.01	5.31
Ten,	3.70	4.10	4.73	5.32	5.84	6.23	6.61	6.83	7.35	7.92	8.18	6.07
Eleven,	4.01	4.38	4.62	4.79	4.64	4.56	4.09	4.92	5.07	5.65	6.82	4.88
Twelve,	4.80	4.84	5.20	5.51	5.72	6.00	6.13	7.08	7.84	9.47	10.71	6.74
Thirteen,	5.83	6.46	6.24	6.30	6.69	7.16	7.92	8.27	9.51	11.29	12.14	7.92
Fourteen,	4.93	5.63	7.39	8.25	9.32	10.13	10.89	12.23	12.93	14.00	16.03	10.07
Fifteen,	8.54	9.03	9.86	11.33	12.27	12.97	13.86	13.64	14.36	14.09	14.57	12.19
Sixteen,	12.25	12.80	16.62	15.84	15.03	15.33	13.79	12.85	11.94	10.96	8.60	13.91
Seventeen,	8.25	10.90	6.36	5.95	5.61	5.19	7.29	8.54	7.08	8.37	8.40	6.48

TABLE 15.— *Showing Annual Growth in Inches of Boston School-girls irrespective of Nationality.*

AGE.	VALUES AT THE FOLLOWING PERCENTILE GRADES.											Average.
	5 Per Cent.	10 Per Cent.	20 Per Cent.	30 Per Cent.	40 Per Cent.	50 Per Cent.	60 Per Cent.	70 Per Cent.	80 Per Cent.	90 Per Cent.	95 Per Cent.	
Six,	2.03	2.02	2.01	2.00	2.00	1.97	1.95	2.00	2.09	2.16	2.13	2.06
Seven,	2.11	2.02	2.04	2.25	2.19	2.23	2.29	2.32	2.34	2.24	2.34	2.17
Eight,	1.67	1.79	1.99	1.83	1.95	2.02	2.09	2.11	2.19	2.42	2.32	2.06
Nine,	1.73	1.65	1.73	1.88	1.92	1.89	1.85	1.81	1.76	1.77	1.78	1.79
Ten,	1.68	1.81	1.82	1.85	1.87	1.92	1.96	2.00	2.09	2.25	2.51	1.97
Eleven,	1.84	1.90	1.86	1.88	1.85	1.95	2.12	2.24	2.35	2.47	2.30	2.08
Twelve,	1.92	1.97	2.10	2.19	2.42	2.47	2.43	2.49	2.63	2.68	2.85	2.46
Thirteen,	2.36	2.26	2.30	2.47	2.46	2.52	2.60	2.53	2.36	2.06	1.99	2.29
Fourteen,	2.26	2.48	2.36	2.10	1.98	1.80	1.52	1.37	1.19	1.20	1.25	1.78
Fifteen,	1.52	1.31	1.22	1.20	1.08	1.01	1.02	1.04	1.21	1.16	0.95	1.16
Sixteen,	0.43	0.35	0.39	0.39	0.49	0.57	0.64	0.63	0.53	0.62	0.59	0.49
Seventeen,	0.40	0.44	0.23	0.39	0.42	0.40	0.31	0.26	0.15	0.11	0.26	0.33

TABLE 16. — *Showing Annual Increase in Pounds of Boston School-girls irrespective of Nationality.*

AGE.	VALUES AT THE FOLLOWING PERCENTILE GRADES.											Average.
	5 Per Cent.	10 Per Cent.	20 Per Cent.	30 Per Cent.	40 Per Cent.	50 Per Cent.	60 Per Cent.	70 Per Cent.	80 Per Cent.	90 Per Cent.	95 Per Cent.	
Six,	2.97	2.82	3.22	3.00	3.24	3.48	3.70	3.91	4.12	4.35	4.48	3.62
Seven,	3.62	3.51	3.83	3.94	4.05	4.19	4.34	4.56	4.89	5.58	5.20	4.18
Eight,	3.57	3.61	3.98	3.96	4.43	4.15	4.74	5.08	4.79	4.96	5.04	4.58
Nine,	3.55	3.96	4.31	4.89	4.60	5.17	4.83	5.05	5.33	5.51	7.36	5.03
Ten,	4.10	3.98	4.10	4.42	4.56	4.78	5.42	5.92	6.46	7.60	6.93	5.28
Eleven,	3.82	3.97	4.35	4.78	5.36	6.22	6.78	7.12	8.04	10.28	12.38	6.40
Twelve,	5.61	6.81	7.71	8.22	8.61	9.46	10.02	10.68	11.16	11.00	13.46	9.47
Thirteen,	6.06	7.40	7.84	8.60	10.39	10.04	11.48	12.42	13.89	13.76	12.69	10.34
Fourteen,	10.82	0.99	10.60	10.81	10.08	9.73	9.25	9.22	7.82	8.66	9.31	9.78
Fifteen,	6.82	8.44	9.67	9.45	8.49	7.36	7.20	6.71	7.70	7.02	7.49	7.65
Sixteen,	5.08	4.71	3.62	3.01	4.79	6.16	7.36	7.56	7.15	0.85	3.72	5.95
Seventeen,	4.73	4.24	5.27	4.07	2.74	1.40	1.41	2.80	2.75	4.89	5.21	3.50

The conclusions to be drawn from these tables will be most readily understood by an examination of the curves on plates 6 and 7, which have been constructed from them, the curves

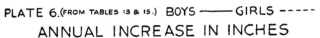

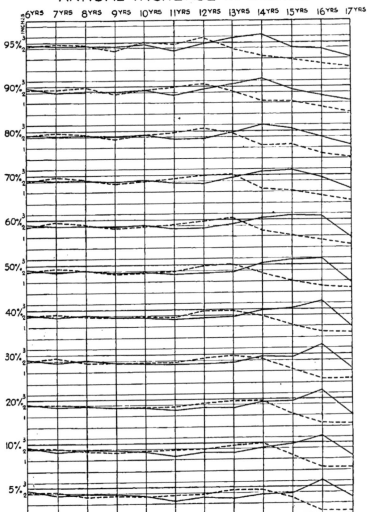

PLATE 6.(FROM TABLES 13 & 15.) BOYS ——— GIRLS - - - - -

ANNUAL INCREASE IN INCHES

representing the yearly growth of the two sexes being, for easier comparison, plotted on the same system of co-ordinates.

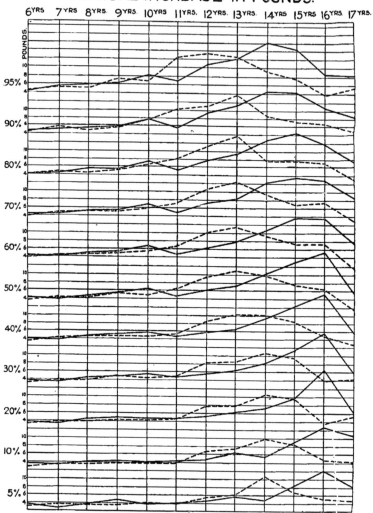

PLATE 7 (FROM TABLES 14 & 16) BOYS——— GIRLS------
ANNUAL INCREASE IN POUNDS.

The following are the most obvious conclusions : —

1. The maximum yearly growth in both height and weight is at all percentile grades greater in boys than in girls, and occurs in boys two or three years later than in girls.

2. The age at which this maximum yearly growth in height and weight is reached is, in both sexes, earlier at the higher than at the lower percentile grades, the range being from twelve to fourteen years for girls and from fourteen to sixteen years for boys. In other words, large children make their most rapid growth at an earlier age than small ones.

3. The curves representing the annual growth of boys are characterized on either side of the maximum by a steeper rise and fall in the lower than in the higher percentile grades, though the maximum itself may be quite as high in the former as in the latter grades. This indicates that the above-mentioned period of accelerated growth in large boys differs from that in small boys rather in duration than in intensity. In girls a difference of this sort does not seem to exist.

4. In boys at eleven years of age there is a period of remarkably slow growth both in height and weight, the curves of annual increase in nearly all the percentile grades reaching at this age a lower point than for several years preceding or subsequent to this age. In girls a similar but less marked period of retarded growth in height is to be noticed at nine years of age, but the rate of growth in weight does not seem to suffer a corresponding diminution.*

One of the conclusions reached in the original article to which reference has been made was that " at about thirteen or fourteen years girls in this community are, during more than two years, both taller and heavier than boys at the same age, though before and after that period the reverse is the

---

* It is interesting, however, to notice that in the curves constructed by Dr. Stevenson (See " Lancet," Sept. 22, 1888) from English and American statistics, and representing the annual increase in weight of " boys and girls of the English-speaking races," the period of retarded growth is a marked phenomenon in both sexes, occurring in boys at eleven and in girls at nine years of age.

See also " Axel Key, Die. Pubertätsentwickelung." (Verhandlungen des X internationalen medicinschen Congresses, Berlin, 1890. Bd. 1., p. 67) This observer finds that in Sweden the period of least increase in height and weight occurs at ten years for boys and nine years for girls.

case." The dependence of this phenomenon upon the fact that the prepubertal period of accelerated growth occurs earlier in girls than in boys was also pointed out. It will be interesting now to inquire in what way this period of female superiority is affected by the fact that the maximum rate of growth is reached earlier in the higher than in the lower percentile grades. The influence of this circumstance is readily understood from an inspection of the curves on plates 8 and 9, which have been constructed from the figures on plates 1, 4, 7 and 10. Here the absolute heights and weights of boys and girls of the same percentile grade are plotted on the same system of co-ordinates, while the ordinates for the successive percentile grades differ from each other by five inches or twenty-five pounds respectively.

In this way the curves are brought vertically over one another and a comparison between them is facilitated. The points where the curves of growth of the two sexes intersect each other are joined by dotted lines, in order that the periods of female superiority at the various percentile grades may be readily compared with each other. A glance at the curves suffices to show that the period of female superiority is to be observed at all percentile grades both in height and weight, and, moreover, that it both begins and ends earlier in the higher than in the lower percentile grades. Thus, in the ninety-five percentile grade, girls begin to exceed boys in height at the age of ten years four months and are in turn surpassed by them at thirteen years, while in the five percentile grade girls do not surpass boys in height till they are eleven years five months old and boys do not regain their superiority till they are fifteen years four months of age. It will be also noticed that the duration of the period of female superiority varies quite regularly with the percentile grade, but that the variations in height and weight are in opposite directions. This is indicated by the fact that the above-mentioned dotted lines converge upward in the curves of height and downward in those of weight. In other words, we may say that, when growing children of both sexes are compared together by corresponding percentile grades, the period of female superiority in height is less conspicuous in tall than in short children, while the period

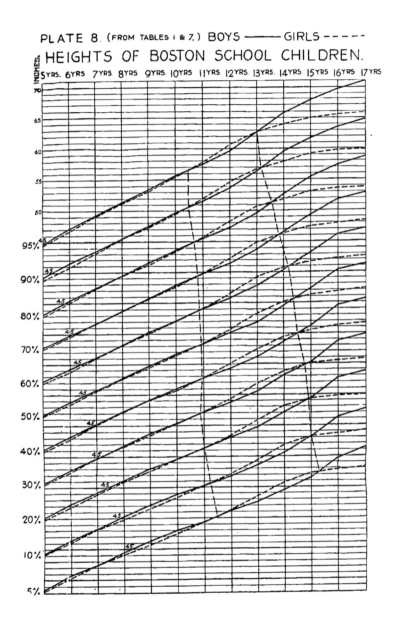

PLATE 8. (FROM TABLES 1 & 7.) BOYS ———— GIRLS – – – –

HEIGHTS OF BOSTON SCHOOL CHILDREN.

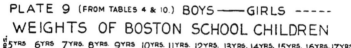

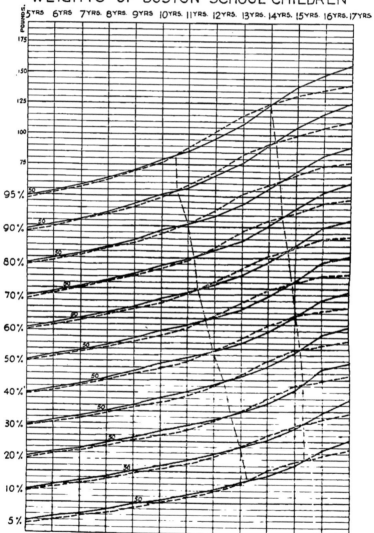

of female superiority in weight is a more marked phenomenon in heavy than in light children.

Among the advantages of this method of discussing anthropometrical results may be mentioned the facility which it
affords for comparing the rates of growth of children of
different nationalities by determining the percentile rank
of the average children of one nationality referred to those
of another nationality as a standard. We may take, for
instance, the observations of Pagliani* on Italian children,
and those of Erismann† on the employees in Russian fac-
tories, and calculate the percentile rank of the children at
successive ages when referred to Boston children as a
standard. The result of this calculation is given in the
following table : —

TABLE 17. — *Showing the Percentile Rank of Italian and Russian Children compared with those of the Boston Public Schools.*

AGE AT LAST BIRTHDAY.	PERCENTILE RANK.			
	ITALIAN (PAGLIANI).		RUSSIAN (ERISMANN).	
	Boys.	Girls.	Boys.	Girls.
Five, ...	below 5	below 5	-	-
Six, . .	5.6	below 5	-	-
Seven, . .	22.1	9.2	75.9	80.7
Eight, . .	26.5	15.8	56.6	63.4
Nine, . .	31.4	29.1	48.9	76.4
Ten, .	20.0	28.0	40.6	51.9
Eleven, .	16.4	25.5	42.5	48.8
Twelve, .	16.1	24.1	36.6	39.0
Thirteen, . .	21.7	23.7	28.7	26.9
Fourteen, . .	21.2	30.0	26.5	22.8
Fifteen, . .	23.7	29.5	29.1	21.4
Sixteen, . .	16.2	32.4	17.7	23.4
Seventeen, ...	13.1	32.2	18.6	22.0
Eighteen, ...	6.6	34.3	15.0	23.6

An examination of this table shows that Italian children of
both sexes are, in early life, very much smaller than Boston

---

* Lo Sviluppo Umano, p. 37.
† Untersuchungen über die körperliche Entwickelung der Fabrikarbeiter in Zen-
tralrussland, Tübingen, 1889. A very thorough investigation based upon measurements of over 100,000 individuals.

children of the same age, and, though they afterwards increase in relative size, they never reach a higher percentile than 31.4 for boys and 32.4 for girls.

The Russian children show in general, with increasing age, a progressive diminution in percentile rank which is probably to be accounted for by the fact that during the earlier period of life only children who are unusually well developed physically are likely to find their way into manufactories. The children from seven to twelve years of age are therefore to some extent selected cases and do not represent the average development of the working population.

It will be noticed that throughout this article it has been assumed that the changes from year to year in the values of the height and weight at the various percentile grades represent the rate of growth of large, small and medium-sized children respectively. This assumption may be criticized on the ground that the values at the various percentile grades do not represent the average measurements of particular groups of growing children but are merely limiting values, on either side of which lie certain percentages of the total number of observations on children of a certain age. To determine how much importance is to be attached to this objection it will be necessary to inquire within what limits the percentile rank of a growing child may under normal circumstances vary from year to year, for it is obvious that if growing children remain practically in the same percentile grade during the whole period of adolescence a comparison of the values at the various percentile grades in successive years will, to all intents and purposes, show the annual increase in height and weight of groups of children belonging in and about those percentile grades, i.e., it will really give us the rate of growth of large, small and medium-sized children. Now it is evident that a close maintenance of a given percentile rank by a growing child is by no means a universal rule, for it is a matter of common observation that very small babies sometimes grow up into very large men and women. Such cases, however, always attract attention from their obviously exceptional character and indicate that, as a rule, there is a certain degree of correspondence between the size of the child and that of the adult.

General impressions with regard to such questions are, of course, of very little value, and before any definite conclusion can be reached it will be necessary to collect large numbers of observations on growing children of both sexes, each individual being measured in successive years or, still better, several times each year, and the percentile rank at each age determined. Such determinations may be made by means of tables 1–10, or by the curves on plates 2–5 constructed from them, but in practice it will be found more convenient to make use of such curves as those on plates 10–13, which have been constructed * with a view to this special purpose.

In these plates, which have been constructed only from the tables which present the total number of observations irrespective of nationality, the age in years and months is given on the sides, the percentile rank at the top and bottom, while the curved lines traversing the plate represent successive inches or pounds. The use of the curves will be best understood by an example. Let us suppose, for instance, that a boy ten years five months old measures fifty inches in height, and it is desired to ascertain his percentile rank. On Plate 10 the horizontal line corresponding to ten years five months is to be followed to its point of intersection with the fifty-inch curve. This point of intersection will be found to lie on the vertical line corresponding to twenty-five per cent. This means that the boy is taller than twenty-five per cent. and shorter than seventy-five per cent. of the boys of his age. A height of fifty-one inches at the same age would give a percentile rank of forty-one per cent., fifty-two inches fifty-seven per cent., fifty-three inches seventy-four per cent., etc.

If we were in possession of a few hundred sets of observations on growing children, each child being measured and weighed annually or semiannually during the period of adolescence and his percentile rank determined in the

---

* Such curves may be constructed by calculating for each age the percentile rank corresponding to each inch or pound by interpolation in tables such as Nos. 1–12, or, still better, from tables such as Nos. 4–15 of the original article, by direct addition to the percentages corresponding to the successive inches or pounds.

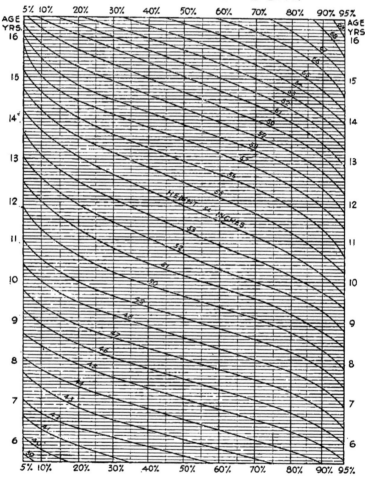

PLATE 10

SHOWING PERCENTILE RANK OF BOYS
OF GIVEN AGE AND HEIGHT.

PLATE II.

SHOWING PERCENTILE RANK OF BOYS
OF GIVEN AGE AND WEIGHT.

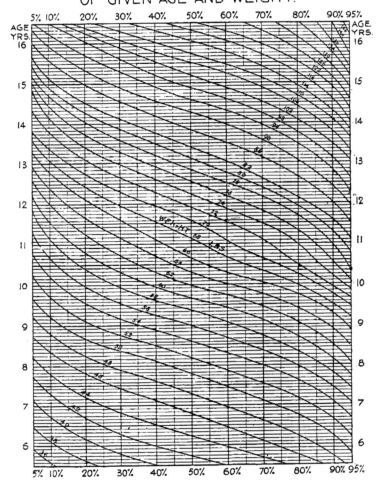

PLATE 12.

SHOWING PERCENTILE RANK OF GIRLS
OF GIVEN AGE AND HEIGHT.

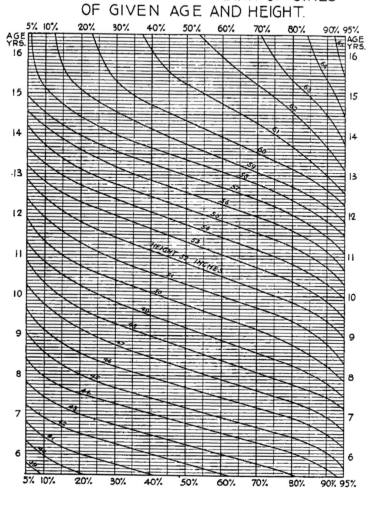

PLATE 13.

SHOWING PERCENTILE RANK OF GIRLS OF GIVEN AGE AND WEIGHT:

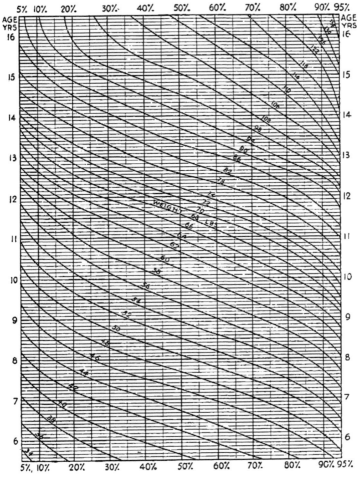

above manner, we should be able to draw fairly accurate conclusions as to the normal range of variation in percentile rank during the period of growth, and to determine how far the rate of growth in the earlier years of life is to be regarded as an indication of the size to be subsequently attained. Unfortunately records of this sort have been rarely kept and still more rarely published. Those which are accessible to the writer represent in most cases the result of observations upon children above the ninety-five per cent. grade in both height and weight, *i.e.*, of a size at which the above tables do not permit us to determine the percentile rank with any accuracy. As an example of the kind of record which it is important to secure the following table is presented, showing the percentile rank in height and weight of two growing girls from the ages of six to fifteen years : —

TABLE 18. *Showing Absolute Height and Weight and Percentile Rank in Height and Weight of two Girls* E *and* F *at Various Ages from Six to Fifteen Years.*

E.								F.							
AGE.		HEIGHT.		AGE.		WEIGHT.		AGE.		HEIGHT.		AGE.		WEIGHT.	
Years.	Months.	Inches.	Per Cent. Rank.	Years.	Months.	Pounds.	Per Cent. Rank.	Years.	Months.	Inches.	Per Cent. Rank.	Years.	Months.	Pounds.	Per Cent. Rank.
6	3½	45.3	89.4	6	3½	50.0	92.7	6	0	44.8	90.0	6	0	48.2	90.5
7	3½	47.6	88.1	7	3½	54.3	89.3	7	0	47.2	89.5	7	0	53.6	91.8
8	3½	49.3	82.6	8	3½	59.8	91.0	8	0	49.1	86.7	8	0	57.3	88.0
9	3½	51.2	83.8	9	6	67.3	92.0	9	0	50.8	85.5	9	0	61.1	84.2
10	3½	53.0	81.7	10	6	73.6	89.7	10	0	52.8	85.0	10	0	67.2	84.6
11	3½	55.8	85.6	11	6	84.5	90.8	11	0	55.2	86.3	11	0	77.8	88.6
12	3½	58.5	85.2	12	6	97.7	92.	12	1	57.7	89.0	12	0	88.4	88.2
13	4	61.0	85.5	13	6	110.9	91.7	12	9	59.6	84.3	13	0	100.1	85.5
14	1	61.8	83.0	14	6	129.2	95+	13	9	62.2	90.8	14	0	114.2	90.2
15	1	62.7	80.0	15	5	134.9	95+	-	-	-	-	-	-	-	-

These records are, of course, not numerous enough to justify any general conclusions, but they are interesting as showing that the percentile rank of healthy growing children may, during adolescence, vary within a range of four or five per cent. on either side of an average value. How much wider the variation may be without passing the limits of health is a question for the determination of which a very large number of observations is necessary, and it is to the public schools that we must again look for the data which shall make it possible to give a definite answer to this and other questions relating to the phenomena of growth. Meanwhile, the above-described variations during adolescence of the height and weight of children at the various percentile grades must be regarded as representing only in a general way the rate of growth of large and small children respectively.

The importance of taking periodical measurements of pupils in the public schools has frequently been urged. In fact, this branch of anthropometry stands in such close relation to physical training that it may be regarded as the test to which systems of physical training must be subjected in order to judge of their comparative efficiency. No teacher at the present day is satisfied to give instruction in any department of learning without testing its results by periodical examinations of the pupils. In the same way the director of physical training can have no certainty that his efforts are well directed unless he can convince himself, by periodical determinations of height, weight, chest girth, strength, etc., that his pupils are making satisfactory progress in physical development.

Here the question at once arises : What amount of progress is to be regarded as satisfactory ? and the importance of establishing a normal standard of development becomes apparent. A rough approximation to such a standard of development in height and weight for the pupils of the Boston public schools has been given in the above tables and curves. By their means it is possible to ascertain whether a given pupil retains his rank (relative to height and weight) among his comrades during the period of adolescence. It is obvious, however, that much more valuable results could be obtained if we were in possession of observations numerous enough to justify the construction of separate tables and curves for children of different nationalities ; for it has been shown that children of American parentage in our public schools are, at nearly all ages, taller and heavier than those of other nationalities. It is also of great importance that some simple tests of strength should be applied to growing children in order to establish a standard of power as well as of size. When a system of annual physical measurements shall have been introduced into our public schools and recognized as of equal importance with the annual examinations in the various studies, we shall be in a position to formulate the laws of growth with much greater accuracy than is at present possible.